Contents

M000015294

Acknowledgements

I am grateful for the help provided by the following: Graham Black at *Earthmovers* magazine; Bomag GB; CNH Group; Doosan Infracore UK; JCB Excavators; Kubota UK; Liebherr GB; Wirtgen Group.

Picture Credits

(5) Bomag UK, *(6)* Case New Holland Group, *(12)* Doosan Infracore, *(15)* Wirtgen Group, *(22)* JCB, *(28)* Kubota UK, *(29)* Liebherr UK, *(30)* Liebherr UK, *(35)* Case New Holland Group, *(43)* Volvo Construction.

All other pictures are by the author

Know
Your
Diggers

Paul
Argent

Old Pond Publishing

First published 2011

Copyright © Paul Argent, 2011

The moral rights of the author in this book have been asserted

ISBN 978-1-906853-81-5

A catalogue record for this book is available from the British Library

Published by

Old Pond Publishing Ltd
Dencora Business Centre
36 White House Road
Ipswich IP1 5LT
United Kingdom

www.oldpond.com

Book design by Liz Whatling
Printed and bound in China

Foreword

I have been interested in construction plant and machinery since I was a young boy and have been lucky enough that for much of my working life I have been able not only to be around them, but also to operate them occasionally.

My aim with this book is to show some of the different types of plant and machinery that are operating on construction sites, scrap yards and landfill sites around the country. Although the title is *Know Your Diggers*, the content is not solely concentrated on earthmoving equipment; instead, it encompasses everything from rollers to high-reach scrap handlers.

I have tried to focus more on the modern machinery that is seen on site these days and to feature manufacturers from across the globe. Although most of the companies featured are still trading, there are, sadly, one or two which are no longer with us.

For each item of plant, I have also tried to include the main place of manufacture even though some are manufactured in various plants around the globe.

The fact is that there are so many different types of machinery from so many manufacturers that attempting to cover everything from everyone would require much more than a single publication. However, I hope that these examples whet your appetite and make you want to know more about the earthmoving machines at work all around us.

PAUL ARGENT

Barford SXR6000

Parent Company:
Invictas Group

Manufactured in:
Grantham, Lincolnshire, UK

Class of Machine:
Forward-tipping dumper

Description

The forward-tipping dumper is an item of plant machinery that can be seen on almost any site. Barford, which is part of Invictas Group, have a long history in the manufacture of construction equipment, producing everything from road rollers to mixers to drainage implements.

Ranging in capacity from 2 to 10 tonnes, the Barford range incorporates swivel skip, high tip and forward tip designs. All of the models in the range have four-wheel drive as standard.

The SXR6000 pictured has a capacity of 6 tonnes with a heaped capacity of 3 cu m. It comes with four-wheel drive, has a turning circle of 12.6 metres and a swivel skip enabling the operator to tip the skip's contents 90 degrees to the left or right of the dumper.

Bell B50D

Parent Company:
Bell Equipment Group

Manufactured in:
Richards Bay, South Africa

Class of Machine:
Articulated dump truck

Description

In 1998, South African manufacturer Bell set up a UK operation to sell their range of articulated dump trucks. Their ability to deliver good service from the start made Bell a big player not only in the UK but Europe-wide. Manufacture of Bell equipment originally began in the late 1940s when the company's prime interest was engineering; however, the first Bell articulated dump truck was not made until 1985.

Bell trucks range from the B20D – a 20 tonne capacity 6x4 version – to the B50D – a 50 tonne capacity truck. Currently, the company promises their customers 'the lowest cost per tonne moved' on all Bell mark VI truck versions.

The B50D is the largest truck in the range and currently the largest-capacity hauler in the UK. Powered by a Mercedes Benz OM502LA V8 engine capable of developing 523 hp, the B50D opposite was photographed at a site in northern England moving 50 tonnes of material to the nearby tip.

BG Pavers C17

Parent Company:
BG Pavers

Manufactured in:
Preston, UK

Class of Machine:
Paver

Description

The name of Barber Greene is synonymous with the laying of tarmac. Since the first machines were produced in 1941, BG has been a constant player in the UK road construction market. They produce a range of pavers capable of laying between 1 m and 12 m of tarmac. Both wheeled and tracked variants are able to lay up to 760 tonnes per hour.

The hot tarmac is tipped into the hopper at the front of the machine and is transferred to the screed at the rear of the machine via a conveyor. The adjustable screed at the rear is fitted with an auger that rotates to distribute the tarmac to a level surface.

The BGP C17 is pictured at the Hillhead Quarry show in Derbyshire. The machine is track-mounted and weighs 13.5 tonnes. Powered by a 188 hp John Deere engine, the machine is capable of laying up to 720 tonnes of tarmac per hour.

Bomag BW120AD-4

Parent Company:
Fayat Group

Manufactured in:
Boppard, Germany

Class of Machine:
Roller

Description

Bomag, part of the Fayat Group, are renowned as world leaders in the compaction field. The company offers a wide range of products from tampers and vibratory plates to tandem and pneumatic rollers.

The range comprises nineteen models that progress from 1.3 to 13.7 tonnes and have operating widths ranging from 800 mm to 2.1 m. The Bomag tandem rollers offer full hydrostatic control giving the operator precise control over all compaction operations.

The BW120AD-4 pictured is a 1.2 m wide, 33 hp, 2.6 tonne roller capable of compacting a variety of materials. It has a vibrating frequency between 55 and 75 hertz.

Bomag BC972RB

Parent Company:
Fayat Group

Manufactured in:
Boppard, Germany

Class of Machine:
Landfill compactor

Description

Not only are Bomag world-renowned for their range of compaction equipment, they also produce a range of top-selling landfill compactors.

Similar in design to wheeled loading shovels, the landfill compactors are essentially heavier, better protected versions. The landfill compactors are fitted with spiked steel wheels that compact the rubbish down, eliminating air pockets and ensuring that all available space is used. The compactors are usually fitted with a dozer blade which is used to spread out the landfill material before compacting.

Landfill sites present fire hazards, so most compactors are fitted with fire suppression systems and all openings to the engine compartment are sealed off.

The Bomag BC972RB weighs 46 tonnes, the second largest of six machines ranging from 21 to 55 tonnes.

Case WX165

Parent Company:
Case New Holland (CNH)

Manufactured in:
San Mauro, Italy

Class of Machine:
Wheeled excavator

Description

From the 1970s to the 1980s, French manufacturer Poclain – often described as the inventor of the wheeled excavator – had the lion's share of sales in this category worldwide. After they bought the company, Case still used the old Poclain machines as a basis for their wheeled excavators. Other manufacturers' machines surpassed them in technological advancements and operator comfort.

Recently, Case has upgraded its range of wheeled excavators to bring them in line with their popular crawler range and to regain lost sales in the wheeled market. The current line-up of wheeled excavators spans from the 10 tonne WX95 through to the 24.5 tonne WX240.

The WX165 sits in the middle of the range and is the most popular Case model. Weighing in at 18.5 tonnes this machine can be equipped with a variety of undercarriage, boom and dipper configurations and has bucket capacities ranging from 0.25 to 1.05 cu m.

Description

Caterpillar M318DMH

Parent Company:
Caterpillar Inc

Manufactured in:
Gosselies, Belgium

Class of Machine:
Wheeled excavator

Built alongside the European specification wheeled excavators in Gosselies, Belgium, the M318DMH is a dedicated material-handling machine. It comes with a host of extras to choose from, including different undercarriage options, cab riser options and boom and dipper combinations.

Power for the M318DMH comes from a Caterpillar C6.6 Acert engine that develops 169 hp. This model is equipped with Caterpillar's 'Smartboom' feature which enhances operator comfort and job efficiency.

The M318DMH pictured is equipped with four-point hydraulic outriggers and a hydraulically elevated cabin. The machine is fitted with a straight boom, drop-nosed dipper and a sorting grab suitable for working in a recycling facility.

Caterpillar 973

Description

Once a common sight throughout the UK, the crawler loader is now confined to specialist tasks, including demolition, working in the steel industry and on landfill sites. However, the versatility of the crawler loader remains as strong as it was in its heyday.

The number of manufacturers producing crawler loaders has diminished so much that Caterpillar and Liebherr are now the main players left in the market. Liebherr produces crawler loaders for John Deere to sell in the USA.

Manufactured from 1982, the Caterpillar 973 is the largest tracked loader produced by the Peoria-based company. This example was being used as a utility machine around a steel works reclamation site. Powered by a Cat 3306 engine capable of developing 210 hp, the big Cat is equipped with a 4-in-1 bucket which makes the 22 tonne machine even more versatile.

Parent Company:
Caterpillar Inc

Manufactured in:
Grenoble, France

Class of Machine:
Crawler loader

Caterpillar M313D

Parent Company:
Caterpillar Inc

Manufactured in:
Gosselies, Belgium

Class of Machine:
Wheeled excavator

Description

The European range of Caterpillar excavators are seen as a premium product in the industry.

The Caterpillar wheeled excavator range has come a long way from the original design, which was based on the German Eder machines. The wheeled excavators are now identical in appearance to the rest of the excavator range.

The M313D is a 13 tonne machine that sits at the bottom of an eight-model line-up which includes the 32 tonne M325MH – the heaviest in the range. Sporting a Cat C4.4 engine capable of developing 125 hp, the Caterpillar M313D is a very powerful little machine compared with others in its weight class. As with other machines in this wheeled excavator range, the list of options is long with variable geometry and offset booms, different length dippers and various outrigger and blade combinations.

The M313D pictured is working on a night-time rail closure in Scotland, ferrying buckets of grouting material.

Description

Caterpillar 966H

Parent Company:
Caterpillar Inc

Manufactured in:
Gosselies, Belgium

Class of Machine:
Wheeled loader

With a long pedigree in the manufacture of wheeled loaders, Caterpillar has possibly the widest range of these machines available. Standard bucket capacities for Cat machines tend to measure as small as 0.9 cu m on the 906H and as big as 14 cu m on the 195 tonne 994F.

More recently, upgraded models in the range have benefited from improved styling that incorporates sloping bonnets which enable the operator to enjoy a better view to the rear of the machine. Some of the larger loaders in the range also offer 'Stick Control System' which provides control of the steering and transmission from a joystick mounted to the left of the operator.

The 966H is a mid-range loader that weighs between 23 and 27 tonnes with a bucket capacity of 3.5 to 4.8 cu m.

This particular example is loading rail trucks with capping soil for a landfill site in Buckinghamshire.

Caterpillar D6R

Description

Caterpillar has a history stretching back as far as the early 1900s when the company was in its Holt days. The first tractors carrying the Caterpillar name were introduced in 1921 and the first D6 variant in 1941.

The current range of European-specification crawler tractors is built in Peoria, USA; these models include the 92 hp D4K and the top-of-the-range 935 hp D11T. All models are fitted with Caterpillar engines developed to combine power with fuel economy.

The D6R bulldozer is a mid-range machine that has now been replaced by newer models. The new D6 can be specified as a K or T variant. The difference being that the D6K is fitted with a standard track frame while the D6T is built with Caterpillar's patented Hi-Drive system.

The D6R opposite is being used to level soils on an opencast mining project.

Parent Company:
Caterpillar Inc

Manufactured in:
Peoria, Illinois, USA

Class of Machine:
Crawler tractor

Doosan DT160

Parent Company:
Doosan Infracore

Manufactured in:
Frameries, Belgium

Class of Machine:
Telescopic handler

Description

Doosan are perhaps best known for their excavators and wheeled loaders; the company name is not usually associated with telescopic handlers. As happens in many business takeovers, Doosan acquired their range of tele-handlers through their recent purchase of the company Bobcat. The Korean manufacturer already had counterbalance forklifts within their product range but the purchase of Bobcat presented a chance to further expand into a new market.

While most manufacturers of tele-handlers build machines with capacities up to 5 tonnes, Doosan beat their competitors hands down with the DT160, which has a lift capacity of up to 16 tonnes at a lift height of up to 10 m. Offering impressive lifting figures the DT160 isn't even the largest model in the range, that accolade is awarded to the 21 tonne capacity DT210.

Description

Fiat Allis FL7B

Parent Company:
Fiat Group, now CNH

Manufactured in:
Italy

Class of Machine:
Crawler loader

Fiat Allis was once a major force in crawler loaders. The company was created through a merger between Italian company Fiat and American manufacturer Allis-Chalmers.

In its time, Fiat Allis produced a wide range of construction equipment from excavators through to pipe layers. The crawler loader range in particular included the range-topping FL20 as well as the mid-range FL7B.

Manufactured in Italy in 1990, the 80 hp FL7B is fitted with a multi-purpose, 4-in-1 bucket and is being used to strip topsoil on a housing project.

Hamm HD90

Parent Company:
Wirtgen Group

Manufactured in:
Tirschenreuth, Germany

Class of Machine:
Road roller

Description

Hamm is a division of the Wirtgen Group which specialises in the production of machines for road building, construction and crushing.

The Hamm HD90 is a mid-range 11 tonne roller used predominately for the compaction of tarmac on road construction jobs. HD90 models are usually fitted with a pair of steel rollers front and rear but can also be specified with pneumatic tyres on the rear.

Inside the cab, the driver's seat, control panel and steering wheel are capable of rotating 180 degrees, thereby giving the operator an all-round view when operating the vehicle in reverse.

The featured machine has an asphalt cutter to the side of the front drum. This enables a joint line to be cut in freshly laid tar providing the paver with a straight line from which to work.

Hamm HW90

Parent Company:
Wirtgen Group

Manufactured in:
Tirschenreuth, Germany

Class of Machine:
Deadweight roller

Description

Although the basic design for the Hamm HW90 has been around for years, this machine is totally modern. Since its inception, the concept for the 'three-point' or deadweight roller design has hardly changed apart from the method of propulsion. The steam engine that once powered this machine is long gone, replaced by a modern, fuel-efficient diesel engine.

The HW90 has a 4 cylinder diesel engine and motion is provided by hydrostatic motors which power the rear drums. As the name suggests, the deadweight roller uses its mass to compact asphalt, as opposed to the more common method of using vibrations. The HW90 has a large operator cabin that offers a good view of both the front and rear of the working area.

The machine opposite is operating on a German road resurfacing project.

Description

Hitachi EG70R

Parent Company:
Hitachi Construction Machinery

Manufactured in:
Tsuchiura Works, Japan

Class of Machine:
Tracked dumper

With conventional wheeled dumper capacities on the increase, the advent of the tracked dumper – which offers the ability to carry similar capacities over more arduous terrain – has been welcomed.

Hitachi has recently launched a 7 tonne capacity tracked dumper, the EG70R, onto the UK market. Unlike the majority of wheeled variants which feature ROPS bars, the Hitachi features a fully enclosed cabin offering a dry, warm working environment. The large rubber tracks with fully oscillating bogies offer excellent traction and flotation over very soft ground.

The EG70R opposite is transferring materials down an old railway cutting. Due to its crawler undercarriage, this machine is able to turn around in very tight spaces; something that a standard wheeled dumper would find difficult.

Hitachi Zaxis 225USRLC

Parent Company:
Hitachi Construction Machinery

Manufactured in:
Tsuchiura Works, Japan

Class of Machine:
Excavator

Description

In June 2000, the Zaxis range of excavators was first launched onto the market and has since been constantly upgraded and improved. Ranging from a 1.1 tonne mini excavator through to a massive 800 tonne mining shovel, Hitachi has one of the most comprehensive ranges of excavators available in the world. European models are manufactured in plants in Japan and the newest Dash 3 models are manufactured for American machinery manufacturer John Deere.

The Zaxis 225USRLC-3 is a 24 tonne reduced tail-swing excavator which sits in the lower end of the Hitachi excavator range. Powered by an Isuzu 4 litre engine which develops 164 hp, the reduced tail-swing machine has nearly 800 mm less overhang to the rear compared to the equivalent Zaxis 210 machine.

Photographed at a landfill site this ZX225USRLC is loading a small fleet of tracked dumpers at the bottom of a spoil heap.

Hitachi Zaxis 280LC

Parent Company:
Hitachi Construction Machinery

Manufactured in:
Tsuchiura Works, Japan

Class of Machine:
Excavator

Description

As with the smaller Hitachi ZX225, the ZX280LC is manufactured in Japan and weighs in at a shade over 28 tonnes.

The excavator is powered by a 4 cylinder Isuzu engine that is capable of developing 188 hp. It has bucket capacities ranging between 0.91 and 1.38 cu m. Unlike most excavators in this weight range this Zaxis 280 is fitted with a hydraulically adjustable three-piece boom. This option allows the operator either to have a reach equivalent to the standard fixed boom or to bring the boom in closer to the tracks for trenching works.

This particular machine has been fitted with a quick hitch and a sorting grapple and is seen here separating concrete from recyclable steel.

Huddig 1260B

Parent Company:
Huddig AB

Manufactured in:
Hudiksvall, Sweden

Class of Machine:
Backhoe loader

Description

These Swedish Huddig backhoe loaders with four equal-sized wheels look like most other large backhoe loaders, but this is where the similarity ends because the Huddig machines have an articulated frame design. The company offers a wide range of options and attachments to make their machines hugely versatile.

Boasting Cummins engines up to 6.7 litres with maximum power outputs of 152 hp, the 12.8 tonne 1260B is in a class of its own. Standard digging equipment comes in the form of a centrally mounted backhoe with a knuckle between the boom and dipper. Options available for the 1260B include rail-road wheels, knuckle boom cranes, pole planters and a variety of other attachments front and rear.

This particular machine has had its wheels replaced with a rubber-track drive system.

JCB 3CX

Parent Company:
JC Bamford (JCB) Excavators Ltd

Manufactured in:
Rocester, Staffordshire, UK

Class of Machine:
Backhoe loader

Description

Since its launch in the early 1980s, the JCB 3CX has evolved into the biggest selling backhoe loader in the world.

The 3CX range is offered with an 85, 92 or 100 hp engine and a choice of specifications up to the top-of-the-range Contractor model. One model, known as the Super Sitemaster, is available with equal-sized tyres all round, while the most popular model sold is the Sitemaster. Standard equipment on the Sitemaster includes four-wheel drive, 4-in-1 front bucket with fold-over forks and an extending rear dipper capable of digging to a depth of 5.97 m.

The 3CX in the photo is a 2007 Contractor model fitted with a Tourquelock transmission that offers up to a 25% saving in fuel use. This UK company also has production lines in India and the USA.

Description

JCB JS360LC

Parent Company:
JCB Excavators Ltd

Manufactured in:
Uttoxeter, Staffordshire, UK

Class of Machine:
Crawler excavator

JCB has a pedigree in excavator manufacturing that dates back to the JCB 7 from 1965 and the popular 80 series from the mid-1970s. After their alliance with Japanese manufacturer Sumitomo in the 1980s and 1990s, JCB upped their game to keep pace with other manufacturers. From this evolved the JS series, a range of machines that are extremely popular with many operators.

Manufactured in the new JCB Heavy Products factory in Uttoxeter, Staffordshire, the JS360LC is one of the larger crawler excavators produced by the company. Weighing in at 36 tonnes and supplied with a choice of either narrow and long or wide and long undercarriages, the JS360 is popular with civil engineering companies and muck shifters alike.

The JS360LC is pictured at the 2009 Site Equipment Demonstration (SED) at Rockingham. It is carrying a 2.6 cu m bucket and is sitting on 800 mm track shoes.

Description

JCB 190
Skid Steer

Parent Company:
JCB Excavators Ltd

Manufactured in:
Rocester, Staffordshire, UK

Class of Machine:
Skid steer

Relative latecomers to skid steer production, JCB launched their distinctive range and immediately gained a major share of the market. The JCB range of skid steer machines features the unique single loader arm design which allows the operator to enter the cab via a door rather than having to make the usual climb over the loader arms.

Both wheeled and tracked models in this range offer the option of a fully glazed cab ensuring that the operator of the JCB skid steer has a warmer, drier environment compared to that of a standard skid steer.

The JCB range consists of eight different models ranging from the 160 wheeled model through to the 1110 wheeled and tracked machines. The 190 model opposite is powered by a 4.4 litre JCB Dieselmax engine producing 84 hp.

23

Komatsu D61PX

Parent Company:
Komatsu Ltd

Manufactured in:
Ishikawa, Japan

Class of Machine:
Bulldozer

Description

Komatsu's array of bulldozers is vast: ranging from the smallest dozer in the world – the 3.5 tonne, 0.75 cu m, D21A-8 – right through to the world's largest production dozer – the 90 cu m D575-3SD which weighs a staggering 152 tonnes.

The D61 is a mid-range dozer powered by a Komatsu 6 cylinder, 170 hp engine. One of the most popular machines in Europe, the D61 comes in two variants. The first variant is the EX with 2.5 m wide tracks on 0.6 m track pads and the second is the PX with 3 m wide tracks on 0.86 m track pads.

One interesting option for the D61PX is a foldable PAT blade. Both sides of the blade are hinged enabling the blade to be reduced down to a width of 3 m, thus doing away with notifying the authorities when transporting the machine from site to site.

24

Description

Komatsu PC130

Parent Company:
Komatsu Ltd

Manufactured in:
Birtley, County Durham, UK

Class of Machine:
Crawler excavator

The most popular weight range for crawler excavators is the 12–14 tonne range and the Komatsu PC130 is one of the best sellers in this category.

In the mid-90s, the Japanese company set up a manufacturing plant in Birtley, County Durham to produce European-specification machines. They now produce the full range, from 13 to 60 tonnes, with the smaller and larger machines still made in Italy and Japan.

The PC130 opposite is a new Dash 8 model. It is equipped with a fully hydraulic quick hitch. The driver of this machine has turned the bucket around and is using it to deposit concrete around kerb stones.

25

Komatsu PC210LC Long Reach

Parent Company:
Komatsu Ltd

Manufactured in:
Birtley, County Durham, UK

Class of Machine:
Excavator

Description

The Komatsu PC210LC-8 is one of the most popular 21 tonne excavators for sale in the UK and is most commonly found as a standard excavator.

The PC210LC weighs between 21 and 24 tonnes and is powered by a Komatsu engine that is capable of developing 156 hp. In standard specification, it can handle buckets of up to 1.68 cu m.

The long reach equipment fitted to the PC210LC is specifically designed for the maintenance of rivers, drainage ponds and other waterways. The equipment gives the machine a maximum reach of 15 m and a dig depth of 11.5 m. This brand-new Komatsu is being used to clean drainage ditches around a hospital in the north-west and is fitted with a reed-cutting bucket.

26

Komatsu PC450LC

Parent Company:
Komatsu Ltd

Manufactured in:
Birtley, County Durham, UK

Class of Machine:
Excavator

Description

The Komatsu PC450 is a rugged, productive machine specifically designed for the European market. The excavator is fitted with Komatsu's Hydrau-mind system which assists in all operations and ensures that the machine's performance is perfectly matched to the task in hand.

The PC450 is powered by a Komatsu 345 hp engine that is turbocharged and after-cooled to provide high productivity and low fuel consumption. The machine is also fitted with Komtrax, a remote access system that enables the machine's owner to track and monitor the machine at any time.

The PC450LC opposite is a heavy-duty version for extreme operations. It has been photographed on top of a pile of limestone at the biennial Hillhead quarry exhibition where it is awaiting the arrival of a dump truck.

27

Kubota KX080

Parent Company:
Kubota

Manufactured in:
Osaka, Japan

Class of Machine:
Excavator

Description

Kubota are often seen as the pioneers of the mini excavator and they have built up an enviable reputation for reliability and innovation within the industry.

The KX080 is the largest model in the KX range and weighs just over 8 tonnes. Powered by a 65 hp direct injection diesel engine the Kubota offers two-speed tracking and features auto-idling to reduce fuel consumption. The driver is very well catered for in this machine with a spacious cabin featuring air conditioning as standard.

This particular model has been extensively modified by the dealer and includes an Engcon rotating, swivelling hitch as well as LGP tracks, a dozer blade and a winch fitted to the undercarriage.

28

Kubota U10

Parent Company:
Kubota

Manufactured in:
Osaka, Japan

Class of Machine:
Micro excavator

Description

With the ever-increasing need to access smaller and smaller spaces, manufacturers have had to look at building more compact machinery. Kubota is one manufacturer that has elected to provide not one, but two micro excavators.

The K-008 and the U10 both weigh around one tonne. The U10 comes with the option of zero tail-swing. Both machines are powered by a Kubota 3 cylinder, 719 cc engine that is capable of developing 10.5 hp. With their track frames retracted, these machines can fit through an opening just 725 mm wide.

Both machines, despite their size, have an impressive digging envelope, offering a dig depth of nearly 1.8 m and a dump height of over 2 m. The bucket capacity for both machines is 0.025 cu m.

29

Description

Liebherr A900CZW

Parent Company:
Liebherr Group

Manufactured in:
Kirchdorf, Germany

Class of Machine:
Wheeled excavator

Although the German company Liebherr has manufacturing facilities all over the world, all their wheeled models are built in Germany.

The A900CZW excavator is the latest incarnation of the popular road-rail machine that is based on the company's 20 tonne wheeled excavator. This model is used as a self-propelled excavator on rail maintenance contracts. It is fitted with a pair of rail bogies front and rear. Power to the rail bogies comes from the excavator's own wheels which are in contact with the bogies to give the vehicle traction.

This particular photo shows the first A900CZW in the UK as it undergoes certification tests. Fitted with a two-piece boom, the machine also comes with a crew cab as standard. This cab design enables the operator to be able to carry a passenger in the rear.

30

Liebherr LRB 125

Description

Parent Company:
Liebherr Group

Manufactured in:
Nenzing, Austria

Class of Machine:
Piling rig

The piling rig is a very versatile machine which looks somewhat like a cross between an excavator and a crane. It can be fitted with a whole host of attachments meaning it can be easily configured to undertake a variety of roles within foundation construction.

However, the standard use for the piling rig is to place steel sheet or tube piles into the ground either by vibrating them or knocking them in with a hammer attachment. Other piling rigs can be fitted with digging equipment such as casing oscillators and rotary drilling tools. These attachments enable the rig to drill large-diameter holes to varying depths which can then be filled with reinforcement cages and concrete to provide solid foundations for large, heavy constructions.

The Liebherr LRB125 is a 43-46 tonne machine with a mast length up to 17.5 m. It has a high-frequency vibrator to drive piles deep into the riverbed.

31

Liebherr R944C

Parent Company:
Liebherr Group

Manufactured in:
Colmar, France

Class of Machine:
Demolition excavator

Description

Liebherr has a reputation not only for material handling machines but also for producing high-quality excavators designed to tackle the rigours of the demolition industry.

The R944C crawler excavator is available with a range of weights between 38 and 59 tonnes. As an excavator used in the demolition industry, the R944 can be specified with a standard, straight or three-piece high-reach boom. It can demolish buildings as tall as 23 m.

The R944C opposite is a more modest machine with a standard excavator boom and dipper. It has full cab and body protection and has been fitted with a spark arrestor system to enable it to work in petrochemical plants. It also has Liebherr's Liku-fix hydraulic quick hitch which enables the operator to pick up attachments without having to get out of the cab. Equipped with a La-Bounty shear, this R944C is tearing down redundant storage tanks.

32

Liebherr R984C High Rise

Parent Company:
Liebherr Group

Manufactured in:
Colmar, France

Class of Machine:
Material handler

Description

Liebherr has a great reputation in the scrap industry for supplying bespoke, powerful machines that can withstand the dirty, arduous environment.

With a huge range of machines weighing 16 tonnes and up, Liebherr material handlers can be specified as either wheeled or tracked machines with a variety of different boom and dipper configurations, cab heights and attachment options.

The R984C pictured was the very first of its type in the world, put to work in a Liverpool metal merchants to load ships with scrap metal. The added height in the turret, between the tracks and the superstructure, and the long boom and dipper, enable the machine to place a load of scrap directly onto the floor of the ship's hold to prevent damage to the vessel.

33

Manitou MRT1440

Parent Company:
Manitou Group

Manufactured in:
Ancenis Cedex, France

Class of Machine:
Tele-handler

Description

The construction industry has always had a need to access tight construction sites and, as a result, the standard tele-handler has been transformed into a rotary machine capable of sitting in one position and performing a variety of tasks in a 360 degree envelope.

Manitou already produces a wide range of standard tele-handlers and is a leader in the rotary tele-handler market. The company manufactures a range of six machines with capacities ranging between 14 and 30 m lift heights. The rotary capability means the machine is able, in theory, to unload a vehicle and deposit the load without turning a wheel.

The Manitou MRT1440 is a 4 tonne, 14 m model. In the picture opposite, the machine has changed its standard forks for a crane attachment and is lifting steel reinforcements to form a wind turbine base.

34

Mastenbroek 20/18

Parent Company:
J Mastenbroek Ltd

Manufactured in:
Boston, Lincolnshire, UK

Class of Machine:
Trencher

Description

Mastenbroek has been building trenchers in the UK since 1977. The company has a wide range of models encompassing hard rock, drainage and de-watering trenchers; these range in weight from 8 tonnes through to 90 tonnes. Although these machines are not commonly seen on construction projects they can be very productive in the right conditions.

The Mastenbroek 20/18 opposite was photographed on an approach road leading to a new wind farm in North Wales. The trencher was tasked with using its chain attachment to dig a constant-width trench for the power cable that connects the turbines to the electricity grid. The small picks on the chain dig out the ground and the resulting material is sent to the side via the machine's conveyor belt. This process of trenching is very tidy and efficient.

35

New Holland B115B

Parent Company:
CNH

Manufactured in:
Imola, Italy

Class of Machine:
Backhoe loader

Description

The New Holland range was originally born out of the Ford-manufactured range of backhoe loaders which has been incorporated into the global brand that is known as CNH.

In keeping with other manufacturers, New Holland produced a large backhoe loader with four equal-sized wheels. Although not as popular as smaller machines, the large backhoe loaders offer better digging performances and higher power outputs than their smaller counterparts. They often manage to do this while still remaining a relatively compact size.

Weighing up to 9.1 tonnes, the B115B is the largest machine in New Holland's line-up. It offers a maximum digging depth of 5.6 m and a front bucket capacity of 1.15 cu m with a 4-in-1 bucket.

36

Pegson Crusher

Description

Parent Company:
Terex Corporation

Manufactured in:
Dungannon and
Coalville, UK

Class of Machine:
Mobile crusher

In the past, waste materials generated from the demolition of buildings were sent to landfill sites for burying but with landfill taxes escalating and more and more natural resources diminishing, those days are long gone.

Many demolition sites now have a mobile crusher on site in order to process their waste, which is then either re-used or transported to other sites. Crushers work using two plates – one that is fixed and one that is moving – to grind the waste material down.

The Pegson tracked crusher opposite is usually used in a quarrying application where it is able to crush up to 200 tonnes of clean stone per hour.

37

Takeuchi TB250

Parent Company:
Takeuchi Manufacturing Co

Manufactured in:
Nagano, Japan

Class of Machine:
Mini digger

Description

Mini diggers are produced by a huge range of manufacturers and they come in many different shapes and weights. Originally the mini digger was pioneered in Japan; now, however, the machines are used on a huge variety of construction and demolition projects all over the world.

One of the more popular brands is Takeuchi. Their range extends from the 0.8 tonne TB108 to the 14 tonne TB1140.

The 5 tonne TB250 is a mid-range mini digger in the Takeuchi product line. Fitted with a 4 cylinder Yanmar engine, the TB250 offers a maximum digging depth of 3.8 m. With a relatively compact width of 1.8 m, the Takeuchi offers great productivity in a compact package.

38

Terex HR32

Parent Company:
Terex Corporation

Manufactured in:
Langenburg, Germany

Class of Machine:
Mini excavator

Description

Terex entered the mini excavator market through the acquisition of German manufacturer Schaeff. Offering a range of machines weighing between 1.4 tonnes and 13 tonnes, Terex sells the Atlas and Schaeff products alongside its own white Terex machines.

The entire range offers both rubber and steel track variants, while the smaller models – such as the TC16 and TC20 – offer an adjustable undercarriage giving a narrow transport width and a stable working platform.

The HR32 weighs in at 3.2 tonnes and is powered by a Mitsubishi 29 hp engine. The HR32 opposite is being used to clear vegetation ahead of a road building project.

39

Terex PT60

Parent Company:
Terex Corporation

Manufactured in:
Coventry, UK

Class of Machine:
Tracked skid steer

Description

The skid steer loader existed for many years before tracked variants started to appear on the market. While better known for their range of mining equipment and articulated dump trucks, Terex have nonetheless designed and built a tracked skid steer loader with a unique undercarriage. This machine is called the Posi-Track.

Essentially a standard skid steer, the Posi-Track's outstanding feature is the suspension on the undercarriage which ensures that the tracks are kept in contact with even the roughest of ground conditions. By keeping the full length of the tracks in contact with the ground while on the move, the machine is better able to reach high speeds.

The PT60 is a mid-range machine powered by a Perkins diesel engine that develops 60 hp. The skid steer can be fitted with a quick hitch making it possible to utilise a vast range of attachments; this particular machine has been fitted with a dozer blade to clear snow.

40

Thwaites 1.5t

Description

The Thwaites 1.5 tonne capacity dumper may be at the smaller end of the Thwaites range but it is the only dumper that can lift its load, swivel it through 180 degrees and dump it at a height of over 1.7 m.

The compact dumper is powered by a 3 cylinder Yanmar engine which develops 33 hp and is fitted with Kinglink – an addition, pioneered by Thwaites, that provides the truck with superior stability and traction.

The Thwaites 1.5t has sufficient tipping height to be able to load a standard builders' skip.

Parent Company:
Thwaites Ltd

Manufactured in:
Leamington Spa, UK

Class of Machine:
Compact dump truck

41

Thwaites 10t

Parent Company:
Thwaites Ltd

Manufactured in:
Leamington Spa, UK

Class of Machine:
Dump truck

Description

Many regard Thwaites as the original dumper manufacturer because they have been building vehicles since 1945.

Their products range from a 300 kg capacity pedestrian-controlled truck to a 10 tonne capacity forward-tipping truck. Options for these machines include forward tip and power-swivel skips as well as power-shuttle, power-shift, manual or hydrostatic transmissions. For added stability on soft terrain, Thwaites trucks come with flotation tyres as standard.

The 10 tonne capacity dumper is the largest capacity machine in the Thwaites range. It is powered by a Perkins D series diesel engine capable of developing 111 hp. The big Thwaites dump truck is fitted with a forward-tip skip, while the smaller machines in the range are available with a power-swivel option.

42

Volvo A25D

Parent Company:
Volvo Construction Equipment

Manufactured in:
Eskilstuna, Sweden

Class of Machine:
Articulated dump truck

Description

Volvo has been a market leader in the manufacture of articulated dump trucks ever since 1966 when it produced the DR631 – the original Volvo truck. Although the capacity of the modern trucks has increased from the initial 10 tonnes and technology has advanced, the original exterior design can still be seen in the newer machines.

Manufactured at Volvo's Eskilstuna plant, the range of Volvo articulated trucks has capacities between 24 and 39 tonnes with power outputs ranging between 305 and 476 hp. All the trucks are available with all-wheel drive enabling them to traverse the most difficult of terrain with ease. They also come with a spacious cab equipped with air conditioning and a passenger seat.

The A25D pictured is the smallest of the 6x6 ADTs in the Volvo range; this model has now been superseded by the A25E. The ADT opposite is being used to haul stone for the construction of a wind farm in North Wales.

43

Volvo BL61

Parent Company:
Volvo Construction Equipment

Manufactured in:
Wroclaw, Poland

Class of Machine:
Backhoe loader

Description

The small range of Volvo backhoe loaders were specially developed in-house and have become popular machines that are capable of pushing the more established manufacturers for sales.

The original design of the BL61 was constructed around the 83 hp Volvo D5D engine. This backhoe loader is the smaller of two models in the Volvo range and it shares many components and features with the larger BL71. The basic specification BL61, which is aimed at hire companies, is available with a 4-in-1 bucket with flip-over forks and an extending rear dipper. A 'Plus' version is also available which provides the option of a 94 hp engine as well as increased breakout forces.

The BL61 Plus is photographed using the extending backhoe to dig out material around a partly demolished building.

44

Volvo EC18C

Parent Company:
Volvo Construction Equipment

Manufactured in:
Belley, France

Class of Machine:
Mini excavator

Description

Volvo was a relative latecomer to the mini excavator market and it was really due to the takeover of French mini excavator manufacturer Pel-Job in 1985 that the company gained an instant customer base. Pel-Job already produced popular machines but with the financial backing and customer support network provided by Volvo, the mini excavator range grew from strength to strength.

Originally the machines were simply repainted in Volvo colours but subsequent model releases and upgrades bore increasing design input from Volvo themselves. The current C series upgrades offer models ranging between 1.5 and 8.6 tonnes and include several zero-tail-swing models.

The EC18C opposite is a 1.8 tonne machine from the recently upgraded range of Volvo mini excavators.

45

Volvo ECR235CL

Parent Company:
Volvo Construction Equipment

Manufactured in:
Eskilstuna, Sweden

Class of Machine:
Crawler excavator

Description

As well as being a latecomer to mini excavators, Volvo was also late to the main excavator market. Once again, the purchase of other companies – in this case Akerman and then Samsung – have provided Volvo with substantial worldwide sales.

The rise in popularity of larger, reduced tail-swing excavators has seen Volvo introduce the ECR235CL, a 24-26 tonne reduced radius crawler excavator. Powered by a Volvo D6E engine capable of developing 150 hp, this Volvo crawler excavator is fitted with a front-mounted dozer blade. With a 185 mm rear overhang, this machine can easily work in confined spaces between buildings.

This particular ECR235CL is working on a road-widening project on a major English motorway. The lack of rear overhang on the back of this excavator enables site traffic to utilise a free lane to pass the Volvo while it works on the hard shoulder.

Digger Dialect

ADT – Articulated Dump Truck

EX – Standard track design for Komatsu bulldozers

HP – Horsepower – Measurement of the power output of an engine

LC – Long Carriage – Longer and wider crawler undercarriage on tracked excavators

LGP – Low Ground Pressure – longer, wider tracks or tyres to allow work over soft ground

MH – Material Handler

PAT Blade – Power, Angle, Tilt Blade for bulldozers which can be hydraulically moved into varying positions

Power Swivel – Hydraulically rotating skip on dumper truck

PX – Wider, longer track design for Komatsu bulldozers

ROPS – Roll-Over Protective Structure – Safety structure that is built into or outside the cabin

4-in-1 Bucket – Multipurpose bucket for greater efficiency

Boom

Dipper stick

The Anatomy of a Digger

Bucket